Codename: Language Inquisitor

Unmasking the Language of Liars

by

Dale Tunnell, Ph.D.

Copyright © 2020 Dale L. Tunnell, Ph.D.

All rights reserved. No part of this publication may be reproduced, distributed, or transmitted in any form or by any means, including photocopying, recording, or other electronic or mechanical methods, without the prior written permission of the publisher, except in the case of brief quotations embodied in critical reviews and other noncommercial uses permitted by copyright law. For permission requests, write to the publisher, addressed "Attention: Permissions Coordinator," at the address below.

Printed by Western Legends Research LLC, in the United States of America.

First printing edition 2020.

Western Legends Research, LLC.

P.O. Box 5343

Sun City West, AZ 85376

www.westernlegendsresearch.com

Dedication

I dedicate this book to my wife Deborah who always provides reasoning in my life. She, above all others, motivates me to search for the lies in life.

Acknowledgments

I want to acknowledge all those who have helped me with this book. Notwithstanding the trainers and educators who have contributed to my wealth of knowledge, here are others I wish to acknowledge.

First, I want to thank the Dark Prince, Prince Andrew, Duke of York. Without his great informative interview, I would not have been able to identify his deception and confirm his extra-curricular activities.

Next, I must thank Hunter Biden, who, for all his wit and sidewinding, is finally coming under the scrutiny necessary to either clear his name or land him in jail.

And where would we be without the woman who gives the sign of Spock while she is swatting at imaginary insects swirling about her head? Yes, I mean Nancy Pelosi, the Queen of the Monster's Ball.

Her Mini-Me, Adam Schiff, was also extremely helpful in providing more than enough ballast to sink his listing ship.

And who could forget Joe Biden? I think he already did. Never a dull moment with Joe; his filtration unit clogged with sand.

President Donald Trump provided a "huge" lexicon to examine his ego and New York style.

Devin Nunes, Rep. Movita Johnson-Harrell, Dr. Judy Mikovits, and Tara Reade all deserve an honorable mention. They contributed so significantly to what I believe is such a target-rich environment.

Most of all I want to thank all those who were conned by politicians and media personalities as they feign concern over our country's welfare. At the same time, they lie to us, promoting their agendas. You inspired me!

Thank you, all!

Table of Contents

Introduction Why I Do What I Do 1

Chapter I Codename: The Black Cat 3

Chapter II Codename: The Dark Prince 6

Chapter III Codename: The Banker 11

Chapter IV Codename: Mini-Me 17

Chapter V Codename: The Embezzler 21

Chapter VI Codename: The Centurion 24

Chapter VII Codename: The Fly Swatter 27

Chapter VII Codename: The Showman 32

Chapter VIII Codename: The Accuser 36

Chapter IX Codename: The Victim 39

Chapter X Codename: The Grifter 42

Chapter XI A Little Background 48

Chapter XII True or False 53

Chapter XIII Tools of the Trade 62

Conclusion 69

About the Author 72

Codename: Language Inquisitor

Introduction

Why I Do What I Do

Finding the lie is all about seeking clues that provide the analyst with a complete picture of the lie, why it exists, and the liar's desired outcome. A single clue could provide all the evidence needed, but most often, I find that the more evidence I uncover, the more complete the picture. The discovery of only one clue does not provide a significant enough foundation for building an assessment. Trust me when I say, "There is always more to the story."

About 15 years ago, I began seeing declassified government documents that provided substantial evidence of a lie that necessitated our entry into the Vietnam War. It was not that I did not already know this but seeing documents spelling it out was disheartening. Having spent two combat tours in the Republic of Vietnam, I indeed questioned why we sacrificed nearly 60,000 American soldiers for nothing more than someone's agenda. Nothing new, politicians are famous for telling us one thing and doing something else that is self-enhancing. With leaking and lying, information continues to drip from a leaky faucet, and recognizing lies for what they are may not get us the actual truth, but we then know what not to believe.

Since I had such extremely negative feelings about deception, I began working on methods to be more proficient at reaching sustainable conclusions. Over time, I added to the processes to keep up with differing types of media distribution. I included basic handwriting comparison, psychiatric content analysis, and diagnosis, and eventually voice analysis. That covered the bases well and finally combined into one comprehensive methodology called SHIELD Analysis.

A few years ago, I conducted training in psycholinguistics for numerous law enforcement agencies around the U.S. As part of the training program, I offered my services to any attendee who completed the courses. Of course, many of those who attended the training took me up on the offer, and I began receiving witness and suspect statements by the dozens to examine.

Eventually, written statements transitioned to audio and video recordings, and while I transcribed many, I discovered I needed to add some personal skills to handle these recorded statements. I went to Israel and trained in voice analysis technology and eventually became a senior researcher.

I discovered that this new technology known as Layered Voice Analysis was highly effective and, in my opinion, was superior to the polygraph and other lie-detection processes. Combining content analysis, basic handwriting comparison, psycholinguistics, psychiatric content analysis, and this new voice analysis technology into a comprehensive method to assess dialogue, I developed the process of SHIELD Analysis.

Over several years and hundreds of evaluations, I was fortunate enough to gain some international recognition. Today, I am using this process to evaluate emails, Twitter feeds, videos, personal letters, testimony, speeches, and a few dozen more human interaction sources.

The next several chapters include some of the case studies on which I have worked. I think most people hate liars who prey on others. Everyone lies at one time or another. It is human nature. Lying gets under my skin, though, when it hurts others. That is why I do what I do.

Chapter I

Codename: The Black Cat

A few years ago, a detective friend of mine sent me a transcribed statement of a suspect interviewed in a sexual assault in a small town in West Texas.

Someone broke into a woman's house and beat her severely, then raped her. She was still in the hospital but slowly recovering. She was a local elementary school teacher who was not known to have any enemies, and she was well thought of by the community.

The attacker was a young male around 20 years old, but it happened so fast, and due to the dark house, she could not see him. The assault was brutal, and she could give few details. She mentioned that she met a young man earlier in the month, and he asked her out, but the woman refused, telling him she did not know him well enough. He seemed perturbed but was very gracious with her refusal. She did not see him again.

The police searched for the obvious evidence, but it was apparent that the suspect used a condom and wore gloves. No DNA, no latent prints. It seems the attacker left nothing behind, only the damage.

The police tracked down the young man who invited her on a date and interviewed him. The transcribed copy of his statement was what they sent me. In the context of the information, he gave nothing up. The police could find no DNA or fingerprint evidence he had ever been to her house.

Dale Tunnell, Ph.D.

The police asked for a timeline of his activities on the night of the assault. He explained his movements, and he provided no information to suggest he was the perpetrator. But it is always amazing to me that the simplest things will often give someone away if the analyst recognizes patterns and clues when they appear.

While reviewing the suspect's interview, I noticed the detective asked when and where the suspect received numerous scratches on his arm. According to the detective, they were very evenly spaced and narrow. The detective thought this might be evidence of defensive wounds received during the assault.

The suspect told them they were from a cat belonging to a "black" neighbor of his. He said he hated cats, especially "blind" ones, nor did he associate with neighbors who kept them.

I called the detective and asked him where he thought the wounds originated. He told me that the scratches could result from a cat. He also said the suspect could not remember his neighbor's name. They canvassed the neighborhood where he lived and could find no "black" neighbors who owned cats.

I examined the timeline he provided for an alibi. He said he was at a local establishment until closing (confirmed), and he left for home at about 1:45 am. He then said he stopped at a 7-11 around 2:00 am (also confirmed) and then arrived home around 0245 am. He said his neighbor saw him drive into his driveway at about 3:00 am (once again confirmed by the neighbor). Then he provided extreme detail I classified as trivia or minutia for everything he did for the next hour when he allegedly fell asleep on the couch.

When I checked for linguistic velocity, I noticed it was about eight lines per hour from 1:45 am until 3:00 am and then increased substantially until he allegedly fell asleep at around 4:00 am.

This radical change in velocity suggested that this may have been when he committed the crime. The detective informed me that though the victim could not remember exactly, she thought her assault occurred between 2:00 am, and 4:00 am. She could not remember for sure because of the severity of her injuries.

I informed the detective that I believed the crime happened between 3:00 am and 4:00 am. I then asked the detective to go back and ask the victim if she had a "black" cat, and if she remembered ever saying anything to the suspect during the assault.

He returned my call later and informed me that not only did she have a "black" cat, but she also suddenly remembered saying to the suspect, "Please don't hurt my cat, she's blind."

When the detective called the suspect back in for further questioning, he recited our conversation and then said to the suspect, "You're a dummy! Did you not suspect we would find DNA?" "What do you mean," asked the suspect, knowing he wore gloves and a condom. The detective informed him, "We scraped the cat's claws and found your DNA. What do you think about that?" It was all downhill after that. The last thing the suspect said before asking for a lawyer was, "I hate cats, especially black, blind ones!" Yay for SHIELD Analysis!

Dale Tunnell, Ph.D.

Chapter II

Codename: The Dark Prince

https://www.youtube.com/watch?v=QtBS8COhhhM&fbclid=IwA R2GxYy4uQBGG4feiVSD58b9NHUpII0fh89fWTSDY5P4C3hD8 ZGKvpEOB2w

The Duke of York and His Sexual Proclivities

Prince Andrew earned that codename due to his relationship with known pedophile and financier, Jeffrey Epstein. He is the Duke of York and a royal prince.

What I found comical was that his sexual exploits were known to British intelligence services. His dossier describing his antics reached beyond Whitehall right into the chambers of the Queen. The honorable Queen put the boot to him, and he found himself sitting on his laurels in the heather, cut off from the Royal Palace. That was more profound than any sanction he faced with civil litigation or FBI investigations.

Emily Maitlis of the BBC interviewed Prince Andrew on a Newsnight Special. For the first time, he spoke about his relationship with convicted pedophile Jeffrey Epstein and allegations made against him over his conduct.

He spoke to Emily Maitlis about his friendship with Jeffrey Epstein and the allegations against him. In a world exclusive interview, Newsnight's Emily Maitlis interviewed Prince Andrew, the Duke of York, at Buckingham Palace. For the first time, the Duke addressed in his own words the details of his relationship with convicted sex offender Jeffrey Epstein, who took his own life while awaiting trial on sex-trafficking charges.

In 2015, court papers named Prince Andrew as part of a US civil case against Epstein. The Prince, who is the Queen's third child, also answered questions about the allegations made against him by one of Epstein's victims, and he discussed the impact of the scandal on the Royal family and his work.

Emily Maitlis asked numerous questions of him. Without corrections, some of the more important ones were:

- In 2008, he was convicted of soliciting and procuring a minor for prostitution, he was jailed. This is your friend; how did you feel about it.

- He was released in July, within months. By December 2010. You went to stay with him at his New York mansion. Why? Why were you staying with a convicted sex offender?

- That was December of 2010. He threw a party to celebrate his release. And you were invited as a guest of honor.

- Because during that time, those few days, witnesses say they saw many young girls coming and going at the time. There is video footage of Epstein, accompanied by young girls, and you are staying in his house catching up with friends.

➢ Another guest was John Brockman, the literary agent now. He described seeing you there getting a foot massage from a young Russian woman. Did that happen?

➢ July of this year, Epstein was arrested on charges of sex trafficking and abusing dozens of underage girls. One of Epstein's accusers, Virginia Roberts has made allegations against you. She says she met you in 2001. She says she dined with you danced with you at Trump nightclub in London. She went on to have sex with you in a house in Belgravia that belonged to Maxwell, your friend. Your response?

➢ You do not remember meeting her? She says she met you in 2001. She dined with you. She danced with you. You bought her drinks. You are in Trump nightclub in London. And she went on to have sex with you in a house in Belgravia belonging to Ghislaine Maxwell. Do you remember her?

➢ Is it possible that you met Virginia Roberts dined with her, danced with her at Trump's, and had sex with her on another date?

➢ She provided a photo of the two of you together. Your arm was around her waist. You have seen the photo. How do you explain that?

➢ Because in a legal deposition 2015 she said she had sex with you three times, once in a London house when she was trafficked to you in Maxwell's house, once in New York, a month or so later at Epstein's mansion, and once on his private island in a group of seven or eight other girls. She made these claims in a US deposition. Are you saying you do not believe her, she is lying?

➢ For the record. Is there any way you could have had sex with that young woman, or any young woman trafficked by Jeffrey Epstein in any of his residences?

> You have talked about a thick skin. I wonder if you have any sense now of guilt, regret, or shame about any of your behavior in your friendship.

Here are some exciting findings of his responses to those questions previously mentioned.

His recorded interview was fascinating, so I conducted a cursory linguistic analysis on audio portion of Prince Andrew's interview. I found over fifty (that is 50) linguistic sensitivity clusters (explained comprehensively in <u>Decoding Treachery</u>) contained in his answers about his relationships with young girls and Jeffrey Epstein.

Three or more sensitivity clusters are considered suspicious. In Prince Andrew's case, he exhibited a risk factor of 56–58%, which is on the upper side of moderate. That means that only about 42% of his responses were truthful. Regardless, his interview was highly deceptive!

When I examined his emotional levels, it was interesting to note that he demonstrated almost no apathy toward the female victims and spent a lot of energy trying to appear casually annoyed but creatively helpful to the interviewer. He was anything but casual or practical. Out of a total of 1336 recorded interview segments, he exhibited stress, tension, highly stressed, extreme stress, and excessive tension throughout the interview 383 times, not very stable in his emotions. He was a good actor but not convincing.

What was even more relevant was that his responses to the previously listed questions produced "inaccurate," "highly suspected," and "false" classifications a total of 787 times, or approximately 58% of his answers.

All these measurements independently suggest deception. Whether collectively, they represent a pedophile and sexual predator, I will leave that to your judgment.

Prince Andrew may be fearful of being questioned by the FBI, but his own family has taken exception to his suspected extra-curricular activities as well. The amount of information he is concealing is enormous and suggests why the FBI wanted to question him. When you add the video to the mix, his body language is also very revealing. Watch when he provides an affirmative answer and continues to shake his head in the negative. When he is telling you yes, his body is telling you no.

After conducting this evaluation, I came away with a couple of conclusions. First, Prince Andrew is a master of distortion. Yet, he is too arrogant to believe he can use aristocratic arrogance to mask his deception. Both his language and body language were compelling in that Prince Andrew fooled no one. Saying he was deceptive is a mild representation of what he truly is. He is a liar.

Secondly, I would not risk the welfare of any minor female by leaving them in his care. I am not saying the Dark Prince is a sex offender or sexual predator. But there are no instances when I would bet against that possibility.

Chapter III

Codename: The Banker

https://www.youtube.com/watch?v=vfb9FIpobL4

Did Hunter Biden lie during his interview?

In October 2019, I conducted a psycholinguistic evaluation of two of Hunter Biden's responses to the questions asked by ABC correspondent Amy Robach concerning whether he capitalized on his father's position as Vice-President of the United States in foreign business dealings in the Ukraine and China.

In his video interviews with Amy Robach, she asked Hunter Biden numerous questions about his relationship with his dad, his addiction battle, and his involvement with Ukraine and

China. Hunter Biden answered important questions presented to him, and these included:

- Do you think you are a distraction right now to your father's campaign?
- You said no one votes or will vote for or against my dad because of me. Do you still believe that?
- How do you think this will all play out in the history books?
- President Trump says you are in hiding.
- Did you and your father ever discuss (your business dealings)?
- Your dad said, I hope you know what you are doing. What did he think you were doing?
- What were your qualifications to be on the board of Burisma? You did not have any extensive knowledge about natural gas or Ukraine itself, though?
- You gave me the reasons why you are on that board. You did not list the fact that you were the son (of Joe Biden.)
- What role do you think that played?
- You are paid $50,000 a month for your position.
- If your last name were not Biden, do you think you would have been asked to be on the board of Burisma?
- Why did you leave the board in April?
- Do you regret being on the board, to begin with?

Now this is a big one, and she phrased this question demonstrates the protectionist position of Amy Robach.

- Also, on Trump's list of accusations against Hunter Biden was that Hunter flew on Air Force Two with his father during an official government trip to China in 2013.

Leveraging that connection for financial gain in an investment deal with Chinese businessmen, Jonathan Lee, the President has repeatedly said that you received $1.5 billion from China despite no experience and for no apparent reason.

"Obviously, fact checkers have said that is not true. This is literally having no basis in fact, in any way. Have you received any money from non- business dealings?"

Nothing like giving a guy the key to the exit door!

- ➤ Did you talk about China or your deal with China? A 12-hour flight over home?
- ➤ What do you say to people who believe this is exactly why people hate Washington? A vice president's son can make money in countries where he is doing official government (business)?

I evaluated Hunter Biden's interview. Following S.H.I.E.L.D Analysis rules and guidelines, I employed psycholinguistics, content analysis, and audiometric voice analysis.

A critical starting point in the evaluation was to accept everything said by Hunter Biden as accurate. This standard avoids an immediate insertion of bias against the former Vice-President, Joe Biden's son. It allows one to perform a word for word, phrase by phrase examination of the content he presented without following a storyline, in other words, looking at the content, not the context.

In each instance where a psycholinguistic rule or guideline violated the standard, I classified the linguistic segment as a sensitivity cluster. The significance of a sensitivity cluster does not necessarily indicate the telling of a lie or even deception. It merely signifies that there is a substantial reason for more investigation.

The appearance of one or two sensitivity clusters may only indicate confusion, frustration, or even an unrealistic expectation on the part of the speaker. However, three or more clusters suggest the presence of deception, but this may result from some hidden intent, concealment of information, or an avoidance of fact; not necessarily resulting from an outright lie.

SHIELD Analysis processes determined that Hunter Biden's answers resulted in twenty-three (23) sensitivity clusters from both interviews combined. The audiometric analysis identified a risk factor of 40% for his first interview about his relationship with his father. The result was at the higher end of the low-risk sector. The second interview about his business dealings resulted in a 49% risk factor, a higher end of the moderate-risk sector. In both instances, the issue is not what he was truthful about but that 40-49% of his answers presented some deception level. Can one trust responses that are truthful only half the time?

The two main areas of concern were his response to a question as to whether he discussed his efforts with his father. He answered in the negative when there was extreme hesitancy in his answer. His answer demonstrated he discussed his business dealings with then Vice-president Joe Biden. The question to ask might be, "Why is this question so disturbing to the Bidens?" Further investigation is necessary.

The other area of concern was the question of whether he received any money from China. Amy Robach's specific question was, "Have you received any money from non-business dealing?" Here his language indicated an explosively untruthful response when he answered, "**No, not at all. No one says... no. Definitely not 1.5 billion? It's crazy.**"

Holy Cow! Are you kidding me? His answer says that he did not receive $1.5 billion. What he received was an amount less than the $1.5 billion. The introduction of the words "not at all," "no one says," and "definitely not" in his response tells a reverse story. The audiometric output on this answer was a resounding FALSE.

His answer also opened another avenue of further consideration. Could he have received some benefit in addition to money? The obvious answer might be he received no personal funds, but his hedge fund received $1.5 Billion, of which he holds a security interest.

Did he receive money from China? In my opinion, not only did he receive money, but his family also benefited as well, as has been well-documented by now. His deception is compelling, not only by lying to the questions but also by what he is concealing.

He also accepted in a round-about fashion that he would probably not have been appointed to the Board of Barisma Holdings had his father not been Vice-president of the United States. His answer here was deceptive only in that he believes that to be a fact while he downplays his response to, "Maybe!"

While the two previous answers were the most prominent, others call into question Hunter Biden's honesty. Of the twenty-three sensitivity clusters, one demonstrated response latency to allow him time to come up with an appropriate answer. Two statements contradicted each other about his discussion with his father about Barisma Holdings.

Two statements acknowledged he would not have been on the Board had his name not been Biden. Ten comments were global justification statements. These statements blamed his notoriety on others' actions, misinterpretations, conspiracy theories, and complete debunking of accusations. He excused his actions, saying he committed mistakes because he was human (this is a big one because he asked forgiveness for getting caught).

There were two instances of moving to the future tense in preparing for his father's election as President. He avoided disclosing his past actions with Barisma or the Chinese by relying on a canned and pre-planned response about Alice in Wonderland to justify why Rudy Guiliani and President Trump are singling him out. The remaining five responses were justification clauses answering questions with questions when a simple reply was sufficient.

Remember, a sensitivity cluster does not necessarily reflect being untruthful. However, groupings and psycholinguistic configurations of this number strongly suggest Hunter Biden is concealing additional information, thereby making his interview deceptive.

Dale Tunnell, Ph.D.

I did find one issue that was extremely interesting to me. While he is close to his family, Hunter Biden appears to hold some level of resentment toward them and specifically to his father. There is a myriad of psychological possibilities, but the fact that he was in drug rehabilitation over 70 times suggests something buried deep in his psyche that does not allow him to move forward with his life. I wonder if it has something to do with his role as the family banker and cash cow.

Any way I look at Hunter Biden's responses, I come away with two conclusions. First, the Banker is not only arrogantly mistaken in his belief that his smooth demeanor masks his deceit, and second, he is a troubled individual who lies to gain acceptance and cover his deficiencies.

Codename: Language Inquisitor

Chapter IV

Codename: Mini-Me

https://www.youtube.com/watch?v=hZOIClcqRAo

Adam Schiff

After listening to this hack for the past four years about the Russians and his "facts" so often refuted, I hesitated to assess anything he presents. But just to give you an example of someone who could be so deceptive by concealment and misdirection, I decided to take one of his media presentations and evaluate it.

These remarks are an uncorrected portion of his presentation:

Dale Tunnell, Ph.D.

"Our guiding principle. Finally, I think what's presented to us here is really so aptly summed up in what the President's own Chief of Staff had to say, when he informed the country that yes, indeed, they had withheld military aid to get this political investigation. He told us to get over it, to get over it. That is, what the President does, we should just get over it. This is essentially what he was saying that we need to just get used to the idea of a corrupt president and get over it. And so, we will have to decide, given that the evidence of this misconduct is so clear and uncontested, are we prepared to just get over it? Are we prepared to say that henceforth, we must expect from this President and those who follow that there will be a certain amount of corruption in which the national security of the country will be compromised, in which the oath of the office will mean that much less in which the belief in the rule of law in the United States will be that much less? Is that what we're simply to get over or get used to? Well, I for one, don't think we should get over this. I don't think we should get used to this. I don't think that's what the founders of this country had in mind. Indeed, I think that when they prescribed to remedy this kind of conduct by a President of the United States putting his own personal political interests, above the interest of the American people, was exactly why they prescribed a remedy, as extraordinary as the remedy of impeachment. And so, we have a very difficult decision ahead of us to make. And I have every confidence that the Judiciary Committee, in consultation with the entire caucus, and our leadership will not only receive this report, as well as the reports of others, and make a proper determination about whether articles of impeachment are warranted. With that, I'm happy to respond to your questions."

I conducted a language assessment on this presentation and found nine sensitivity clusters in six different areas. I will explain shortly, but most of this speech was deceptive only for missing and exaggerated justifications. Representative Schiff avoided lying by directly shifting the responsibility of determinations of President Trump's alleged abuse of power to other members of his Committee and downplayed his role.

His speech was his attempt to obfuscate information and elevate its importance, in other words, concealment and exaggeration.

The deception was more in line with what he did not say. He did not say there were "High Crimes and Misdemeanors," nor did he accept any role in which fairness was concerned, a lack of commitment. There is definite concealment, but without further investigation, it is impossible to determine the importance currently. But to be clear, he presented information that he knew (or hoped) would be accepted by supporters...he insinuated guilt. He suggested that the public and other members, including the Judiciary Committee, accept his conclusions based on a higher calling and superior responsibility.

His offer of proof could be accepted or refused depending upon ideological belief. He used polarization to support his position and degrade his opposition. There were some indicators that he may be concerned about a lack of support from other party cohorts. His language demonstrated some frustration and uncertainty. So, the bottom line, he was deceptive in what he concealed and how he justified his position, but I was unable to detect an outright lie. The care he took to avoid an appearance of deception by using facts as only he knew them also suggests that the speech was reviewed multiple times and memorized.

The six areas of concern are as follows:

1. Surgical denial: plays word games to deceive someone – answering a question relying on words chosen in specific definition or content, i.e., Exploiting gaps in question to be masked but not explicitly answering the question asked. Skirts the issue. Schiff answered the reporters' questions with non-answers.

2. Verb tenses: changes in tenses from past to present or future – reliving an event or projecting denial of future events but not past events, relying heavily on what President Trump might do in the future.

3. Positional appearance: the numerical importance one places on lists of persons, events, activities, objects, and people in a series of references. The arrival in sequential descriptions – the closer to the speaker in the statement indicates the level

of importance. The farther away means the appearance is less critical and eventually least important, the changing framework of occurrences and past, present, and future relevance.

4. Repeated validation: validation based on some trivial example, document, text, etc. Relevance based on some lofty calling or higher responsibility.

5. Unnecessary words and irrelevant details – trivia and minutia: information or language introduced in a statement which is not necessary to topic of discussion; a person either considers the information relevant to only them or they may be attempting to evade sensitive conversation

6. Using verbal modifiers: leaving a way out of a statement, i.e., "Currently," "rarely," "hardly ever," "basically," "essentially," "seldom," "most of the time," "usually," "sort of." In other words, his speech is an excellent example of an attempt at CMD (cognitive manipulation and deception) but avoiding outright lying. He is terrific at cognitive manipulation and deception (CMD).

Adam Schiff is the worst kind of con artist. He attacks others by exploiting information he knows to be false and elevates himself as the sole purveyor of honesty and integrity. It amazes me that so many Americans fall for his deception.

Yet, it is all a ruse. Mini-Me demonstrates his insecurities by lying about others and setting himself up as the arbiter of truth and virtue. There is nothing virtuous about Adam Schiff. He is a liar, and no one should trust him with their wallets or anything to do with national security information.

Chapter V

Codename: The Embezzler

https://thehill.com/homenews/state-watch/473132-pennsylvania-lawmaker-charged-with-stealing-from-nonprofit

Pennsylvania Law Maker Charged with Stealing from a Non-Profit for the Mentally Ill that She Founded

A Pennsylvania state representative faces charges over allegations that she stole more than half a million dollars from a nonprofit she founded.

Rep. Movita Johnson-Harrell (D) allegedly spent more than $500,000 from Motivations Education and Consultation Associates on personal expenses, including vacations, designer clothing, luxury car payments, real estate purchases, and past-due mortgage payments.

When you read the article, pay close attention to her statement:

Johnson-Harrell issued the following statement to the Associated Press in response: "I am saddened and dismayed by the nature of the allegations brought against me today," She said. "I vigorously dispute many of these allegations, which generally pertain to before I took office, and I intend to accept responsibility for any actions that were inappropriate."

She has reportedly agreed to resign. She remarks, "I am saddened and dismayed by the nature of the allegations brought against me, today." She is telling the reader she regrets discovery.

Her statement said she "vigorously" disputed "many" of these allegations, which "generally" pertain to before I took office, and I "intend" to accept responsibility for "any" actions that were inappropriate. In truth, she does not dispute all the allegations because someone discovered her crimes, and she cannot avoid prosecution. Her remark that she "intend(s)" to accept responsibility means she does not.

While the term "vigorously" could be deemed a strong adverb, the word "many," softens it, which means "not all."

The word "intend" suggests she has not made up her mind yet, and she is attempting to avoid public acceptance if possible. And when she used "any," she will assert that most of her actions were not inappropriate.

There is also a breakdown of rules regarding the proper configuration of a truthful statement. We call this the Ti-Mi-Ti rule or 1st trivial issue, main issue, and 2nd trivial issue. They should be in a range of 20%, 50%, and 30%. In her statement, her configuration was 2 Ti's and no Mi – The first clue of deception.

Also, while it may appear so, there is no admission. Movita described what was important in this statement because it pertains to what law enforcement has already discovered. She accepted "no" responsibility for anything that occurred in the nebulous past.

While the Embezzler may not be lying, her language identifies that she is highly deceptive. She is an embezzler and an opportunist. The fact that she embezzled from the mentally ill makes her particularly loathsome. Oh, and did I mention she was a politician?

There is more, but you get the idea! Psycholinguistics is fun, huh?

Dale Tunnell, Ph.D.

Chapter VI

Codename: The Centurion

https://www.youtube.com/watch?v=csli46lansc&list=PLkj3a8vFPq2v-Vj1kdgYSqwIrS8puAsMZ&index=7&fbclid=IwAR27vRWscyHJimQhbRxDBH1vlbP9b8XON8xMZu1oubYXY4AjYMNLNZ4s82E

DEVIN NUNES: Nunes: My phone records do not match what Schiff, Dems put in report

He lamented about how the House Intelligence Committee had subpoenaed his telephone records unbeknownst to him. He explained this act's relevance and explained why he would be taking legal action against those involved, mainly Adam Schiff.

I evaluated his interview and examined language using SHIELD Analysis. I determined a moderate risk of deception (41%) in his responses, suggesting that he was engaging more in concealment than outright lies. It may be reasonable to conclude that while he stipulated his intention to take legal action, he did not want to tip his hand and was, therefore, somewhat obtuse in his indirect responses.

However, here are a few examples in his direct commentary when he demonstrated a lack of commitment to the truth on several occasions. Use of first-person, future tense such as "would have," or "I would" is not as powerful as the first-person, past tense such as "I did not" or "I didn't" and should be considered a questionable response.

He also began commentary several times and then substituted words, which changed the flow of the explanation. These substitutions are frequently an indication of anxiety and discomfort resulting from the question asked. Another example was answering questions with a follow-on response, "Right," or answering a question with another question.

Other areas to be highlighted included answers to a higher power. Explanations include displays of dismay such as "Lo and behold" and "All of a sudden." There was extensive minutia introduced during several of his answers, revealing his intent to evade the questions. There were also several distinctive changes between his use of "say and said," vs. "tell and told." The shift in usage is a recognizable clue that something caused a change in demeanor from passive to aggressive. Not knowing the answer, that change merely encourages further investigation.

My examination was only cursory, but still, I counted 19 sensitive clusters. Not necessarily a reflection of lying, there is a clear demonstration of deception through concealment and a need for further investigation.

In other words, not everything the Centurion said during this interview was 100% accurate. But then he is a politician, and like so many, he is very practiced at "fading the heat." With political types, one needs not to look for degrees of truth but rather the level of risk that they are untruthful.

Codename: Language Inquisitor

Chapter VII

Codename: The Fly Swatter

https://www.youtube.com/watch?v=ClyMdm7que0&fbclid=IwAR3LF1xL30IJHalvdSJNcsodQfRfDBo4UkQmH9u2IU5JiUbgzPRxD2aMK08

Pelosi tells a reporter, "Don't mess with me" when asked if she hates Trump

Here is my brief assessment of Nancy Pelosi's speech about being Catholic and not hating anyone. I also included her reasoning for the impeachment proceedings and her rant about her feelings about President Trump being a coward.

Dale Tunnell, Ph.D.

At first glance, without any formal evaluation, a person who already has a directed bias and hatred toward the Democrat party would automatically call Nancy a liar and rank her level of humanity a step higher than pond scum.

However, remember from a past assessment I conducted that human lie detectors are only about 51% accurate in spotting liars, and you could flip a coin to reach the same level of chance. So, for the most part, they are not that accurate. This speech is a prime example of that fact. Regardless of what you may think about Nancy Pelosi, not everything you saw in her remarks was representative of what is going on in her mind.

Also, for deception to exist, there must be intent. A lie is an intentional misrepresentation of fact, and the speaker or writer must know that the information they are providing is, in fact, false.

If a person misrepresents a material fact, yet believes that misrepresentation to be factual, then the information does not exist as a lie, merely a misconception. In many instances, people jump to conclusions. Upon hearing what they believe to be false information, they automatically conclude the contributor has lied to them. So is the case with Nancy Pelosi.

An evaluation of her combined speech, including the debate with the reporter, her attestation about her Catholic upbringing, her statement that she does not hate President Trump or anyone else for that matter, and her reasoning for pursuing the impeachment proceedings was about 31% deceptive which places her in the low-risk range.

But not for the reasons you may suspect. I would have expected that range to be much higher, more in the 70-80% range. But I was wrong, and I am confident that was a product of my own bias and distrust of politicians and bureaucrats. Her deception was more in line with concealment of information and intentional misdirection of her animus rather than outright lying.

There were a few sensitivity clusters that suggested concealment, and they indicated some hidden personal emotions. I examined Pelosi's emotional profile (an indicator of where she is, emotionally and mentally) and determined that she was overly emotional yet confident and energetic during this episode. She displayed minimal concentration, stress, passion, and reasoned thought.

Nancy's speech demonstrates that her emotions, not her reasoning, were controlling this episode. This fact was the first sensitivity cluster. The next had to do with her declaration of faith.

She made religious statements as if answering a higher religious or moral authority, and she referred to personal morals or upbringing. Her comments suggested she responded to a higher code of conduct or a member of a particular class of people who no one should question. Piousness is a neon sign in an open-ended statement. These statements usually represent a crutch to support the original message when there is a possibility that the audience might not accept her reasoning. She was concerned about being believed.

Two separate statements regarding her Catholic upbringing as a Christian and her directive to the reporter, "Don't mess with me," constituted two more sensitivity clusters.

These comments are not necessarily a lie but bringing this concept into the mix suggests that she is not disclosing just under the surface having to do with her faith. Confusion? Misdirection? Maybe both. I doubt an MRI would fix that.

She exhibited several speech latencies or gaps between declarations, each time representing another sensitivity cluster: not necessarily deception but a reason for further investigation. There is a possibility that her mental acuity is declining, a normal aging process. Or it could be the cumulative effects of alcohol intake. One thing for sure – Donald Trump is in her head, and she loses sleep over him. Nonetheless, there were about ten or more of these.

When Pelosi stated she does not hate President Trump, she may be telling the truth because she clearly defined her commitment with strong verbal tense. This act is somewhat of misdirection, though, because it is obvious, she views the President as a subordinate human being while she might not hate anyone. But if this is her belief, then she cannot be accused of lying.

The misdirection I spoke of is more important to Republicans than they know. While Pelosi directs some of her attacks at the President, her hatred is more intensely pointing at her party members.

Remember, she warned her compatriots not to go down the impeachment road, and they went against her wishes. Feeling boxed in, she probably thought she had no alternative but to support members of her party. She displays significant emotional pain at losing control over the far left, and she will most likely seek revenge at a convenient time.

Because of her remarks about the President, she is very vindictive and selfish. Nothing unusual there, as this fits most people in powerful positions. However, because she is so confident of her earlier recommendations to avoid the impeachment approach, she seems ready to tell members of her party, after their demise, "I told you so!"

In all, I counted approximately seventeen (17) sensitivity clusters during her speech. Usually, more than four or five indicates deception but might be only in concealment of fact or hidden intent.

It stands to reason for her personality; she intends to seek revenge on those who usurped her authority. More specifically, though, the number of sensitivity clusters suggests a strong basis for more investigation regarding her youth and Catholic upbringing. I sense an undercurrent, and she is swimming against it. There is something buried deeply there and maybe the basis for her discomfort with powerful men like President Trump. That is only a guess on my part and certainly not a clinical diagnosis. But there appears to be a cauldron of hatred buried deeply in her psyche, and it had to come from somewhere.

In summary, the Fly Swatter is highly deceptive. Is she a liar? Not so much. But her skills at masking her intent are highly developed, and if she says she is coming after you, do not underestimate her cunning or misunderstand her comments. I would never be in her company without a panel of witnesses.

Dale Tunnell, Ph.D.

Chapter VII

Codename: The Showman

https://www.youtube.com/watch?v=yzDzUyPRyzg&fbclid=IwAR1dNvNdYmwlFOlKzKQc3gmme9OtWbNvhTyPlwNIwn77CHcwbLHZJ91-VBA

Evaluation of President Donald Trump Interviewed by Jackie Nespral of WTVJ NBC6, Miami

There have been several requests to evaluate the deceptiveness of answers provided in interviews by President Trump. I selected an interview when Jackie Nespral of WTVJ NBC6 Miami Interviewed Donald Trump in Florida on December 7, 2019.

I chose this interview because it was a clear recording, and I could quickly identify deceptive cues. Also, while this interview was not explosive, nor the questions necessarily tricky, any propensity for deception would show up clearly. The results might lead to a more critical conversation.

In other words, if he were deceptive in this easy interview, he would most likely be deceptive in one more sensitive. In general, there were some interesting takeaways from this interview.

First was that his responses fell in the low-risk range of deception, 31%. Any rating below 35% is considered low risk. In the lower ranges of risk, scores are most often affected by cognitive processes such as stress, emotion, concentration, etc.

In this instance, the President was too stressed, rising to the highest score possible. At the opposite end of the spectrum, barely registering, were passion, mental effort and efficiency, low energy, and concentration. I identified mid-range results in emotion, uneasiness, cognition, hesitation, and anticipation. So, what does all this mean?

The lower end scores suggested that this was just another interview for him, and there were no significant challenges. He was not particularly worried about the interview, and he did not expend many mental resources in his answers. The result was low energy, no passion interview requiring only minimal concentration.

His anticipation level suggested that the interview was impromptu, but he responded adequately and saw the interview as an opportunity. His responses were slightly more emotional than logic-based, as are most of his interviews, and he was moderately cautious in answering questions.

Of most concern to me was the extreme stress he demonstrated during the interview. While most of the questions can be considered minimally difficult, those would not have contributed significantly to the high-stress measurement. The high-stress source was his questioning about the effort to impeach and remove him from office.

Nearly one-third of the interview focused on this topic and, not surprisingly, the source of several sensitivity clusters. Sensitivity clusters are groups of anomalies that identify reality changes in speech and do not necessarily suggest deception, but the greater the number, the more probable the degree of information concealment. Information concealment is a fundamental contributor to deception.

In reviewing his language content, I found twelve (12) sensitivity clusters in 6 different categories. The most prominent reflected his exaggerated terms about his Ukrainian telephone call repeating that it was a "perfect call." He frequently engages in this type of reflection using excessive terms of importance.

Another comment referred to his poll numbers "going through the roof," another exaggerated expression reflecting deception. While his numbers are increasing, they are certainly not "through the roof." However, we all know people who exaggerate the significance of everything in their conversation. Most often, their speech is not an intent to deceive but rather their style of presentation. In my opinion, his exaggerated responses are more style related.

Another area of significance is his statement, "we" did nothing wrong." This remark is critical because the only one accused of a wrongful act is President Trump. One would expect him to use the pronoun "I" rather than "we." So, I would question who President Trump refers to; President Zelensky and him or those present during the call with President Trump? His comment produces a sensitivity cluster and a basis for more examination.

There was also a point in his discussion when he discussed Nancy Pelosi without naming her. Usually, there is an introduction of a personality by name or position before engaging in a debate about them. It is a minor issue but still a sensitivity cluster.

Finally, I believe the President's elevated stress stemmed from the topic of Robert Levinson. His stress level spiked during his statement about the welfare of Mr. Levinson. Words such as, "He may be alive," "He wasn't very well, even long ago," signals received, "he may be alive, he wasn't very well," suggests an attempt to provide some hope for the family but setting everyone up for a terrible ending.

He then changed directions and diverted responsibility to President Obama for not getting him back in exchange for Iran's billions of dollars. He transfers liability in the event Levinson is determined to be deceased. This comment may be the one area of prominent deception when he stated, "We are working very hard," at getting Levinson back.

In my opinion, the configuration of President Trump's statement suggests he already knows that Mr. Levinson will not be coming home alive and that singular exchange contributed the most to the high level of stress exhibited.

I believe that President Trump is an exaggerator to the extreme. While his comments most often do not rise to the level of a lie, they often contain concealment. The exaggerations stem from his upbringing as a construction boss and developer in New York, and I attribute those to his style. I might add, it appears he inherited this style from his father.

However, when the Showman conceals information, I find that the mere act of concealment does not necessarily stem from malintent. He is probably not looking to cause injury, but his deceptive intent is clear. He has other plans he is not about to disclose. And while many refuse to believe it, he is most frequently the smartest person in the room.

Dale Tunnell, Ph.D.

Chapter VIII

Codename: The Accuser

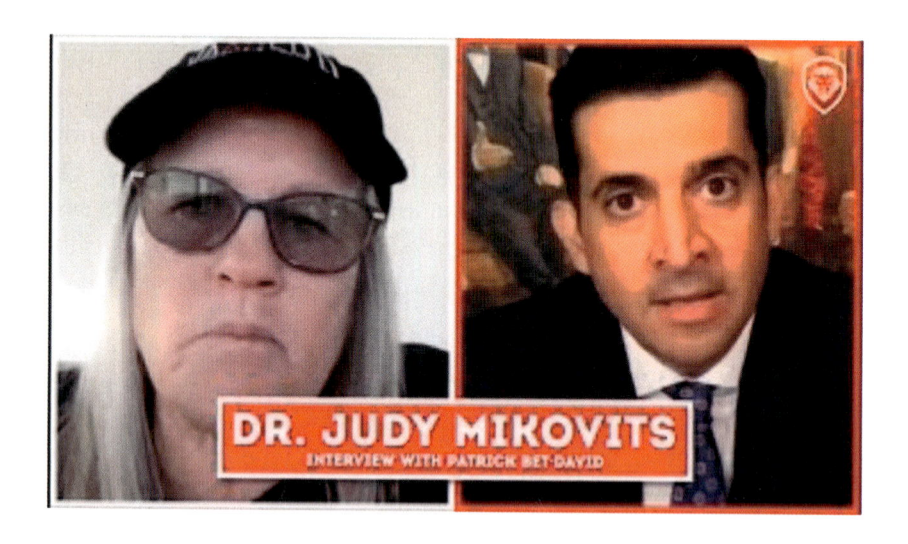

https://coronanews123.wordpress.com/2020/11/19/dr-judy-mikovits-claims-of-50-million-americans-dead-of-covid-vaccine-not-unsubstantiated/

Dr. Judy Mikovits Accuses Dr. Anthony Fauci of Corruption and Conspiracy

Patrick Bet-David interviewed Dr. Mikovits on April 29, 2020. During the interview, she accused Dr. Anthony Fauci of corruption and conspiracy in the mishandling of the COVID-19 pandemic. Both Facebook and Youtube immediately took down the interview video for violating "community standards" and presenting information contrary to established pandemic protocols. While the video was difficult to obtain, I eventually did so, and I scrubbed the audio portion of her presentation for both psycholinguistic and audiometric analysis.

It is my opinion that Dr. Mikovits was not lying. However, that only means she was not intentionally attempting to mislead anyone. If a person believes in what they are saying, then by definition, their information is not the product of a lie.

In Dr. Mikovits's case, she believed what she was saying. Her declaration relates to her presenting information that she based on supposition and the joining of unrelated facts. In this instant, while she believes her story, she also knows that the facts as she presents them are disjointed but assembled to support her accusations. That is generally how conspiracy theories work. She is aware that this is the case, and subsequently, she is deceptive.

The evaluation of her emotional profile shows that she was only minimally distressed while relating her story. She had several instances of hesitation, repeated words (as many as four repetitions of a single word) several times, so there was some anxiety. Yet Mikovits was not emotional. Her levels of confidence - 88%; energy – 94%; passion – 100%; and concentration – 70%, all pointed to her sincere belief in the individual details she presented, just not the aggregation of those details. Her deceptive risk factor hovered around 49%, medium risk. This score suggested that her deception was based more on the strength of her belief and knowing she was assembling unrelated facts to support her proposal rather than concealment or outright lying.

The Accuser's story is real problem is that it is based more on her opinion than on substantive facts. An investigation of each of the events would require more in-depth scrutiny to determine their validity as she described them. At this stage, she is making accusations without relevant proof. No one should take her testimony about Dr. Anthony Fauci as gospel. At the same time, no one should discount it. Unfortunately, I cannot determine the accuracy of her accusations without a comprehensive investigation. However, at this stage, I would not only take her word for it.

However, considering all the drama about Dr. Fauci, there may be more to her story than just supposition.

Chapter IX

Codename: The Victim

https://www.youtube.com/watch?v=4dBdfUudsIk

Assessment of Tara Reade's Allegation of Sexual Assault Against Joe Biden

This case intrigued me, so I spent some extra time with her statement. It probably will not mean much to you, but I was initially concerned about her speech latencies.

Latencies consist of hesitations and opportunities to plug in the information that might be irrelevant or made up. It turned out these were indices of high stress, high concentration, and heightened anticipation before truthful utterances. These are standard traits of sexual abuse victims. Then I heard her utter a phrase she attributed to Biden, "come on, man!" One of his signature phrases, it has a distinctive meaning when he gets frustrated.

There was a possibility that she infused this comment to make her story sound more believable. However, just to be safe, I conducted an audiometric and linguistic assessment to see if I could sort that out. I will not go into details because it would take about ten pages to write up, but not only is she being truthful, the psychological damage done to her is both extreme, recurring, and very apparent.

She was victimized not only by Biden but also the media, the "Me-Too" movement, and those politicians who pledge to support victims of sexual abuse but not so much. They frequently talk out of both sides of their mouths.

I go at this kind of issue from an unbiased starting point, so I gave both Biden and Reade neutral positions. I make no conclusions as to what led to this encounter nor what resulted afterward. In my opinion, the audiometric evidence of the incident is conclusive.

After I conducted this exam, I can say unequivocally, her story is true. Unfortunately, there is no corroborating evidence, and the case may be limited by the statute of limitations, thereby negating any possibility of prosecution.

My educated guess is this is not the first time she has been a victim, but it was undoubtedly the most prominent.

To be clear, while my linguistic assessments are subjective and run about 87% correct, the audiometric program I use is the best in the world and immensely powerful.

While several people on Facebook will undoubtedly disagree with my assessment and attack me for saying something I can very aggressively defend, they should take the emotion out of the equation and pay attention to the evaluation. They should also recognize that several media types, including Krystal Ball, Megan

Kelly, and Mika Brzezinski, are engaging in sensationalism despite their bringing it to the public's attention. The Victim is once again victimized, but this time by media pundits.

Like the Boy Scouts, the political environment, the Catholic Church, Hollywood, the media, and corporate America are frequent breeding grounds for predators. In many cases, they go unchecked. Unfortunately, Reade's story is not unusual in today's environment, as evidenced by the latest media accounts.

Dale Tunnell, Ph.D.

Chapter X

Codename: The Grifter

https://www.youtube.com/watch?v=seu_C08yAAM

Full Interview: Biden Denies Sexual Assault Allegation from Tara Reade | Morning Joe | MSNBC May 1, 2020

Recently, Vice President Joe Biden was interviewed by Mika Brzezinski from MSNBC about the allegations against him of sexual assault on Tara Reade. The interview has gone viral, and numerous video clips are available on Youtube.com and other media sites. I conducted my evaluation using the methods I developed involving psycholinguistic and audiometric calculations.

What I learned was confusing at first, but after several reviews of my findings, the picture is clear.

Let me begin with the evaluation I conducted recently of Tara Reade's recorded statement of her encounter with Joe Biden. I concluded that she was truthful when she alleged Joe Biden sexually assaulted her. I believe he also treated her with utter disdain and ultimately caused her to experience more psychological trauma when he relegated her to the level of an unimportant urchin with no credibility. Added to her diminished self-esteem because of Biden's influence, the Me-Too-Movement's support mechanism failed her at the time she reported and continues to do so.

I performed another evaluation from Joe Biden's interview using the same methods I used with Tara Reade's recorded statement. In general, I concluded that Joe Biden was deceptive in many of his responses. However, because there was statistical strength among many of his answers, indicating both truth and deception, I determined two different themes of importance when he provided answers to Brzezinski's direct questioning. First, did Biden sexually assault Tara Reade? And second, was he truthful about any supported recordation that might exist, either at the National Archives or the University of Delaware.

I have seen this as one of the most interesting in nearly 20 years of performing evaluations on audio and verbal conversations. You will see why in a moment. There are several conflicting forces at work here. The first is the intent to deceive or conceal information. The second reflects the mental status of a subject. The third has to do with cognitive impairment, and the fourth relates to embarrassment. There are many other issues, but these are the categories I focused on because they were the most prominent to me.

While I have a Ph.D. in Psychology, I am not a licensed psychologist, and therefore it would be unethical to diagnose Joe Biden. I do not wish to imply that suggested findings resulted from personal diagnosis. The actual mental assessment of Joe Biden requires a therapeutic review by someone licensed to do so. Subsequently, I can only rely on results as they relate to his speech and language, a skill of which I am fully versed.

Dale Tunnell, Ph.D.

Concerning Tara Reade's claim, it is my opinion that Joe Biden was deceptive when he stated he did not sexually assault Tara Reade. But cognitive impairment masked many of his answers regarding the assault. He was not truthful concerning any records that may support her claim, but he may not have lied about the sexual assault. So, you may ask how this can be; neither individual is lying about their experiences. Confusing, huh?

To sort out this dilemma, I had to apply one other function involving mental capacity. What I found was disconcerting. Tara Reade was truthful in her representation of the assault. Her description of the events and attending emotional factors suggested that her portrayal of the act and follow-on emotional impact was in line with what one would expect from a sexual assault victim. Both psycholinguistic and audiometric assessment confirmed that finding.

For Joe Biden, I found two discrepancies (there were others, but these were the most obvious). First, Joe exhibited fear and embarrassment during several questioning segments. Both emotions related only to the possible existence of documents that may be detrimental to him and those probably located within the University of Delaware archives.

His fear was not just some concern that there may be something of public importance, but an extreme worry that something might exist that would cause him embarrassment and embroil him in a criminal investigation. The deception occurred at this point, and it was both evident and prevalent. The embarrassment factor presented several times when Mika asked him about how his fellow Democrats treated him and the media vs. Judge Kavanaugh's treatment. He again exhibited embarrassment when Mika described the physical act of the assault. That embarrassment factor might be expected just by listening to the allegation, but it could be that flashes of memory of the sexual assault were presenting and interrupting his delivery.

On to the assault allegation, Joe Biden was deceptive, but he may not have been lying in his response that he did not assault Tara Reade. If he believed he was innocent of sexual assault or did not remember the episode, he cannot be lying by definition. His

44

deception may have arisen from related incidents. Regardless, I wanted to investigate his mental acuity directly from his language.

I applied a process called Psychological Content Analysis and Diagnosis. It is a software application, highly validated, developed by my mentor and friends, Louis Gottschalk, M.D., Ph.D., and Robert Bechtel, Ph.D.

The software analyzes conversations converted to text and automatically diagnoses potential mental health issues and psychiatric disorders, suggesting potential problem areas for therapeutic planning. The results cannot be tampered with, thereby excluding examiner bias. I use this process to gain psychiatric insight. In this instance, the results were alarming.

The results I found are in line with when the media questioned his mental acuity. I divided Biden's interview into two separate but equal components. From his language assessment of the first component, I believe Joe Biden presents with shame anxiety and highly elevated health-sickness scores, indicating he is preoccupied with his health or sickness. There appears to be a personal concern that his life accomplishments were not as extensive as he planned. In his estimation, his lack of achievement by his standards created a frustration that he was dependent on so many other people to escalate his importance. Not unnatural for a high-profile individual and often leads to a somewhat selfish view of things.

The second component was more definitive and suggested that there may be a transient or more permanent psychopathological process occurring. Biden demonstrated mildly elevated cognitive impairment and continued with highly elevated health-sickness scores. The verbal analysis suggested these readings may be the result of some current events.

Contrary to many beliefs that Joe Biden is an outgoing, friendly, and warm individual, this analysis suggested quite the opposite. His scores indicate that he dislikes cordial and warm relationships with others. He is inclined to be pessimistic and mentally depressed or present with some type of social phobia.

I did not expect to find evidence suggesting a range of psychiatric disorders. I believe an organic factor negatively affects Biden's ability to filter his comments and beliefs. Two of the most recent examples include:

"...we're in a situation where we have put together, and you guys did it for our administration, the President Obama's administration. Before this, we have put together I think the most extensive and inclusive, voter fraud organization in the history of American politics."

"And like I told Barack, if I read something where there's a fundamental disagreement (between he and Kamala Harris) we have based on a moral principle, I'll develop some disease and say I have to resign."

It looks to me like he has already mapped his way out of the oval office and identified his replacement.

Planning of this type may be typical for a politician. Telling us about it is not. He is telling the truth but not intentionally.

Only Joe Biden's neurologist can accurately diagnose any existing mental problems. But the software I use is accurate to the point that I would expect his medical team has already assessed his diminishing mental acuity.

What does any of this have to do with Joe Biden's truthfulness about the allegation of sexual assault? Quite simply, while he probably remembers Tara Reade and the issue of her complaint, he may have masked the memory of the incident itself. Even though this may be disturbing to potential voters and supporters, the situation is not unusual for elderly individuals. It may be that he pushed the episode from his memory and does not wish to re-engage it, thereby setting up a conflict called cognitive dissonance or two opposing thoughts at the same moment. This opposition results in anxiety, embarrassment, and emotional distress.

Before his death, my father believed he rode with John Wayne and Gary Cooper as a participant in many famous movie escapades. Often, he could not remember essential people in his life, expressing confusion and aggravation when questioned about them. Joe Biden may be experiencing similar issues. Only those close to him know the truth, but his condition in no way gives him a pass.

I believe more problems are on the horizon for the Grifter. Like his son Hunter, Joe Biden is a con-artist. He developed his trade over 50 years in government service. While he may have diminished capacity, he is still one of the most deceptive politicians I ever evaluated. Never mind his foibles of incorrect pronunciation or his creepy machinations with females and their hair. Joe Biden is not trustworthy, and anyone who trusts him to place America first is seriously jeopardizing their future.

Dale Tunnell, Ph.D.

Chapter XI

A Little Background

My parents encouraged me to get a good education and keep up my grades through continuous study. In my house, homework always came before play, and my grades showed the results. Both instilled in me a sense of honor and integrity that I carry with me even today. They taught me right from wrong and continuously reinforced that right always triumphs over evil and it is always best, to tell the truth. Unfortunately, as I experienced more in life, I learned that right does not always triumph over evil, and almost no one believes in always telling the truth.

Gradually, the halo of human honor and truthfulness began to fade. I learned that being truthful does not always serve us well. We know from a young age that frequently, telling the truth lands you in hot water. Despite what your parents told you, they would not be angry at you if you always told them the truth. That was a crock! It was more about the deed you were responsible for than the story you told. And your parents telling you that you would not be in trouble if you always told the truth? They lied! You were still in danger. Now the lesson is buried in your psyche forever.

Over time, I developed an innate sense of deception and decided that my God-given talents could protect me against liars. It turned out that this was a false sense of security. My "sixth sense" was not as accurate as I thought it was and often caused me to distrust my capacity to judge others' personalities. These false perceptions are called "false-positives" and "false-negatives." I learned that if I was correct at least 50% of the time, I was doing well. But what about the other half? I could just flip a coin and reach the same odds. Flipping coins was not an effective way to judge truth and deception. I knew I had to find a better way.

All told, I spent 41 years in law enforcement, mostly as a criminal investigator in federal, state, and local departments throughout the Western United States. But while I worked, I went to night school and took weekend classes to attain baccalaureate and graduate degrees in criminal justice and management. Eventually, I earned my doctorate in psychology. At the same time, I gained specialized training from two people: one, a relatively unknown international expert in language analysis and the second, a real pioneer in psychiatry.

The first individual ran a small classroom laboratory at his residence, and he often provided training to law enforcement agencies worldwide. His name is Avinoam Sapir. He was once a polygrapher with an Israeli police agency. He developed training in language analysis that was and still is the foundation of nearly all language analysis training today. He was the first to immerse me in an endeavor that changed my life. My respect for him is unchallenged.

The second individual was Dr. Louis A. Gottschalk, Ph.D., a psychiatrist, and researcher. He was Professor Emeritus at the University of California at Irvine, and they built a psychiatric research institute in his name, The Gottschalk Psychiatric Institute. He gained national prominence by announcing in 1987 that Ronald Reagan had been suffering from the diminished mental ability as early as 1980. He came to this conclusion by using the Gottschalk-Gleser scales, an internationally used diagnostic tool he helped develop for charting impairments in brain function, to measure speech patterns in Reagan's 1980 and 1984 presidential debates.

Louis Gottschalk coinvented software that uncovered a link between childhood attention deficit disorder and adult addiction to alcohol and drugs. In 2004, at age 87, he published his last book, *World War II: Neuropsychiatric Casualties, Out of Sight, Out of*

Mind, when he chronicled his work on Post Traumatic Shock Syndrome (PTSD). Louis was my friend and my mentor. He introduced me to a technique he referred to as "content analysis of psychiatric states." He published innumerable books on the subject, many of which I retained with his endorsement and autograph.

Along with a friend of his and mine, Robert Bechtel, Ph.D., they created and validated a computerized method of evaluating transcribed text to identify psychiatric issues. The internal operating system is the same as Louis used in his assessment of President Ronald Reagan. I employed the same tool in determining the probability that President-elect Joe Biden may have cognitive impairment in the form of dementia or senility along with other psychiatric issues. Look for more of an explanation in my next book, *Codename: The Grifter.*

In 2007, I went back to school in Tel Aviv, Israel, where I attended training with Nemesysco, Ltd., to learn skills in voice analysis technology. The company developed a computerized method called Layered Voice Analysis. The CEO, Amir Liberman, was gracious enough to take me under his wing and invest his time to train me. Over several years and numerous research projects, I attained the status of Levell III Certification – their highest certification as a researcher and instructor. Today, I am one of three international experts in audiometric voice analysis using their technology.

Over time, I attained a specialized level of expertise in psycholinguistics, content analysis, and audiometric voice analysis so I could train others to detect deception. I wanted to teach others that by merely listening to someone or analyzing a document's content, one could recognize a liar and avoid victimization.

What I learned throughout my career was that in nearly every walk of life, people lie. We are bombarded with lies from 10 to 250 times a day, and the worst possible feeling is when one learns they have been used and victimized by a lie.

Lying has consequences but generally for the one deceived. Advertisers, salespeople, media, and politicians often use some form of deception to conceal information. Even more critical, institutions you learned to trust from a young age have become untrustworthy and wield power for the sake of their benefit and not yours. They attempt to modify your behavior by lying to you, causing you to believe that which is not valid. Political parties and their media lackeys are experts at this phenomenon. Take, for example, the COVID-19 pandemic. Did any of these so-called leaders and media pundits follow the science when they mandated and supported lockdowns, facemasks, and social distancing? Or was this just behavior modification of the masses for their benefit?

Victimization occurs because liars use a pervasive tool often challenging to detect, called deception. I wanted to do something about it. You see, the pain of being lied to and being used can be devastating. It became my mission in life to uncover lies and help other people unmask lies as well. You can discover what the liar does not want you to know and, at the same time, learn a skill that over 95% of Americans do not possess.

What I learned from all my training and education is that finding the truth is sometimes impossible. But identifying the lie is crucial to our livelihood, safety, and security.

On a dismal and cloudy afternoon after the Steele Dossier, the DOJ Inspector General's report, and the impeachment debacle, a group of retired investigators and I were sitting in a coffee shop discussing politicians and media hacks. You can imagine how the conversation devolved into personal opinions and pet names for unpopular personalities. It was probably not the best moment to assess the inequalities of humanity. The cynicism was palpable with a turbulent undercurrent. In other words, these guys were just not in a forgiving mood. Public execution of the worst political violators was on the table.

To change the subject, I mentioned the significance of evaluating language to determine the validity of statements. They were all familiar with the language or content analysis, and the discussion evolved into an evaluation of the voice in determining truthful statements. I felt that liars should be recognized for their deceptive intent, and there were legitimate measures to isolate deception.

One individual commented that it should be like the "Inquisition" of the Dark Ages. Drag them out of their offices and have public hangings! While the thought did appeal to me, I suggested that everyone lies at some point regardless of motivation. I said, "We can't hang everyone. There would be no one remaining, including members of this group."

I opined that if one learned simple tools to identify the lie, the truth will eventually surface and provide a protective mechanism against the intentional liar. Yeah, I know. I am not always realistic, but it was the thought that counts! However, during the spirited discussion, one of my friends said to me, "You are like a damned inquisitor. Only you hang the liar with their language. Hey! I'm going to call you the *Language Inquisitor*!" And thus, the nickname.

Without exaggeration, it happened just like that. No, really! Whatever! I found it to be a useful moniker.

By now, you have read the previous evaluations, so I wanted to leave you with some idea of how I examine these dialogues. In the remaining chapters, I will provide insight into factors one must consider before conducting truth and deception evaluations.

Learning the actual definition of a lie is critical because that is the beginning point of the evaluation. When you understand the different types of lies, you can more easily attribute uniqueness to falsehoods you discover. And finally, I believe you should understand the process of evaluation and the tools I use to reach my conclusion.

So, read on and discover some tips and tricks I use to gather evidence of deception. There may be an "Aha!" moment when you realize that you would like to immerse yourself further into the technology of psycholinguistics and audio metrics.

Chapter XII

True or False

Have You Ever...

Have you ever read a letter or a news article that you did not feel right about but could not figure out why it was bothering you?

Did you ever hear someone tell a story that did not seem credible, but you were afraid to say anything for fear of offending them?

Have you ever been in a position where your intuition told you something was wrong about someone, but you did not trust your intuition about their lack of integrity?

Have you ever received a text or email from someone, only to learn later that someone intended to cause you embarrassment or financial harm?

What would it mean to you to know with certainty that someone who pretends to be your best friend is being deceitful behind your back?

Have you ever had a loved one tell you something only to learn later they were distrustful?

What if you could discover the truth before being blindsided?

These situations happen all the time, and frequently the victim is unable to detect deceit.

Does that include you?

What is a Lie?

People the world over consider an untruth as a "lie." However, that is not always the case. A person may say something untruthful, like, "I'll be there shortly," or "I earn about $3000 a month," when neither statement is entirely correct.

Or how about when the wife or girlfriend asks, "Does this dress make me look fat?" Not a comfortable spot to be in if you are a male, you should answer, "You look great, honey," unless, of course, you are completely insane and you are looking for trouble. What you may be thinking versus what you say are two different processes, so you avoid hurt feelings and lie.

The fact is statements of this nature are not so much untrue as they are inaccurate or exaggerated. Technically they are lies, but to what level are they harmful? Here is one, "It was a perfect phone call," a reference by President Trump to the phone call between President Zelinski of Ukraine and him. President Trump may have believed the phone call was perfect, but many others did not, and it became foundational to his impeachment.

However, one might ask the question, "Was it a lie, or was it an inaccurate or exaggerated statement?" There seems to be a subtle distinction between the two, but there is a difference.

Specialists interested in cognitive manipulation and deception (CMD) are more specific in their definition. They define a lie as something that is not only untrue but said for fraud. The statement intends to manipulate your behavior by having you believe something said in a manner to exclude the exposure of the truth.

Consider these Different Types of Lies

Look at different types of lies:

- Jokes
- White lies
- Defensive lies
- Offensive lies
- Embarrassment lies

Jokes

When a person tells a joke, its primary purpose is to entertain. Its communication is usually in jest and is not associated with stress or deception (unless you are a comedian and you are looking for laughs). What is essential is not identifying the joke as a deception but determining the person's emotional range telling the joke. Any ridiculous comment such as, "I just flew in, and my arms are tired," or "My dog thinks he's human," is likely to fall into the category of a joke.

White Lies

How about when you say something to someone to avoid upsetting or hurting them? The primary intent is not to deceive but to prevent embarrassment or emotional pain, as in the two examples above. While this is still a form of deception, the purpose and physical reaction differ compared to other types of lies.

Defensive Lies

Often the most common form of deception, a person may deny a truthful allegation to protect oneself or someone close to them. A parent may accuse a child of spilling milk on the floor. The child responds that they were not the ones who dropped the milk when, in fact, they did. The response is a defensive lie to avoid responsibility for the act and any recrimination or punishment that comes with it.

Offensive Lies

A person tells a vicious lie to gain an advantage that the truth may prevent. This type is less personal and not used to defend against pending recrimination or perceived harm. A deception type often used in advertisement, business, and politics; the minimal expectation of danger is usually diffused and presents little jeopardy to the lie's originator.

Embarrassment Lie

A lie presented to avoid the embarrassment of any circumstance. A prime example is a statement, "I don't know that woman," or "I never had sex with that woman," when the facts say otherwise.

How about, "That is not my child, I don't even know the mother," embarrassment lies presented recently by Hunter Biden when DNA evidence proved otherwise.

What is a Lie Detector?

A lie detector is any tool designed to determine one's level of truthfulness. The basic premise is to monitor involuntary body physiology to identify and analyze a person's state of arousal, fear, and stress. Using individual sensors to monitor heart rate, blood pressure, respiration, and electrical conductivity of the skin, an examiner can detect measurable differences under stress.

Stress is a result of what we call cognitive dissonance. Two thought processes not psychologically consistent with each other, as in truth or deception, triggers discomfort until the speaker resolves the situation. That is why polygraph and voice analysis work. However, due to the need for an examiner's subjective interpretation of the test result, lie detectors are, as a rule, not accepted as evidence in court. The same is true with the program I developed called SHIELD Analysis, which we will discuss later.

The expense associated with acquiring, training, and operating a polygraph machine or voice analysis application is usually prohibitive to the average person. SHIELD Analysis presents a more reasonable approach because the process entails an evaluation of language style and content. The two functions are trainable and significantly less expensive while offering high levels of success. The primary learning curve is relatively easy and can be applied almost immediately upon becoming familiar with the process. However, to gain comprehensive skill requires a commitment of time and practice. Not insurmountable, but to become a practitioner, one must be genuinely interested in the process.

The human brain plays a large part in the generation of speech and writing. The entire thought and speech process co-occur through mindful interpretation, reflecting feelings of confusion, depression, and pain in speech and writing. That is why language is so important. The words used in communication reflects the mental process occurring at the time of the speech or writing. The study of this process is what one calls psycholinguistics.

We often receive information from media sources such as cable news, op-eds in newspapers, blogs, tweets, internet reporting, and other communications outlets. Most people base their opinions on sound bites and short quips of information without any validating process. The lack of integrity in reporting leads to a misinformed population and divisiveness in the current political and social climate. So how do we enhance our understanding of the information provided to us?

Ask yourself these questions:

- How do I sort out comprehensive information from slanted opinions driven by political operatives and discourse?
- How do I identify the misleading and inaccurate communications derived from speeches and testimony?

The answers come from a two-step process.

Step 1:

Know the source of information presented; these are usually danger signs when you see or hear them. There are information sources that will automatically disqualify any info they offer. Never rely on social platforms for your information. The information is generally slanted toward a particular agenda and masqueraded as fact.

Identify the connections between sources and information outlets.

- Who is married to whom?
- Where was the source previously employed?
- Where does the funding for the source originate?
- What are the general relationships which might impact a factual representation?
- Ask who benefits from the information.
- Get some history of the presenter and put the information into some context.
- Recognize that most media output benefits some plan or organization.

Step 2:

When written speech is involved, analyze both the language's content and context using a program like SHIELD Analysis. I do this by examining transcripts and documents containing the written expression. I look specifically for:

- Word linkage
- Subjective meanings of each word used
- Relations among words
- Pronouns
- Commitment of statements
- Sensitivity terms
- Changes in language
- Unnecessary language or information

- Hesitations, distortions, repetitions
- Expressions of emotions
- Self-corrections
- Higher/lower-level vocabulary
- Document velocity and structure

Recorded Speech

When recorded speech is involved, I use several programs to identify cognitive processes that might impact validity. These include:

- Psychological disorders
- Drug use
- Deception
- Intensity
- Conflict
- Latency or speech delay

Whenever possible, I use various programs that will provide me with the most information for evaluation. Quite often, I use electronic analysis programs because:

- The subject need not be present during the analysis.
- Language is non-specific – Analysis focuses on acoustic parameters of the voice linked to cognitive processes and includes a psycholinguistic assessment.

While there are drawbacks to full linguistic and auditory analysis individually, using combinations of these tools always provides me with an insight I might otherwise miss. Using both processes can be time-consuming, and investment in the software applications is expensive, involving $7,000 to $10,000 for software and training.

In the next chapter, I will describe what programs I use and how I use them.

While I Am at It

Did you know that cognitive manipulation is a universal human trait? It occurs in sales, debate, and everyday conversation. But it is also foundational to every lie ever told. Cognitive manipulation is not necessarily lying or even deception. It is merely attempting to change another person's behavior with alternative views. However, deception is the act of causing someone to accept as accurate or valid what is false or invalid by lying or concealing information. And lying is making an untrue statement with intent to deceive. So how do we defend against cognitive manipulation and deception (CMD)? We build a defense strategy utilizing a method or technology to uncover information hidden within CMD. We use the program called SHIELD Analysis.

When someone is telling their story, listen for them to use the first-person singular, past tense. An example is saying," I drove to the store" or "I did not assault that person," etc. When they do so, there is a significant level of commitment. You can generally believe them. "Did not" is much stronger than "would not."

Noncommittal and tentative future tense statements demonstrate no commitment on the speaker's part. If they cannot commit firmly to their account, why should we accept it as accurate? We should not. Did you hear anyone continue using noncommittal and tentative future tense statements such as, I would not, we would not, I would never, during the impeachment hearings? Do not just hear the soundbites. Listen to the language.

Did you know that intentional memory lapses like forgetting significant events or "misremembering"/ remembering trivial information but unable to recall details of a seminal event is an easy signal of deception? Also, denial expressions such as trust me;

believe me; honestly; really; frankly; I am not lying; let me be perfectly honest (frank) with you. Using any of these phrases suggests the opposite!

I just thought I would throw all that in to give you a bit of advantage against liars!

Dale Tunnell, Ph.D.

Chapter XIII

Tools of the Trade

As I have improved my skills in the deception detection business, I concurrently broadened my repertoire of techniques and applications to narrow searches for keywords, linguistic patterns, statistically significant vocalic markers, and other symbols that indicate deception. As I have said in the past, one cannot always learn what the truth may be. But if you can find the lies, the real story writes itself.

I wrote a book called <u>DECODING TREACHERY</u>, which explains the linguistic side of examining a person's language by individual words, phrases, and word patterns. The book provides a complete course of instruction to learn language analysis without using any software program. A person can learn to become autonomically responsive to speech or writing in a relatively short time. The software applications I utilize merely speeds up the process. However, to use any of these aids, a comprehensive understanding of language analysis basics is still necessary.

Let us look at each of the tools I use and examine the sequences I follow to make the most out of each one when looking for deception in written, spoken, and recorded speech. I follow a few rules that help me focus on what will produce the results.

At the same time, these rules keep me from following rabbit trails that use up my resources of concentration, decision-making, and time expenditure. Because locating the falsification is only half the effort. Once I discover the deception, there must be reasoning and explanation to explain the deception's intricacies.

And finally, without some reporting capability, the effort may give me that warm fuzzy feeling, but I cannot share it with others. Writing is the necessary evil, but alas, it is only one medium to share the results. Others include blogging (still requires writing), podcast, video creation, and audio presentation. I use them all at some time or another.

The Jump Start

When I begin my effort, I start with the most straightforward attempt first. I determine the nature of the exemplar. Is it a written document, a face-to-face conversation, or a recorded conversation? Suppose the communication is in the form of a face-to-face conversation. In that case, I can only rely on my skills to immediately assess the integrity of others' information. After doing this for several years, I find that I can usually sort fact from fiction quickly. Combined body language, facial expressions, and linguistic fervor produce a favorable environment to detect a lie.

If I happen to be hearing a speech, it is essential to determine if the speaker is the originator or written by someone else. Written addresses themselves produced by someone other than the speaker provide little reason to examine it because they are not from the speaker but an unknown author. Usually, seeking veracity from something written by someone else offers little justification to waste time evaluating it. However, if the speaker created the document, it might be worthwhile to assess it linguistically.

Regardless, there is definite merit to evaluate the speech using audiometric voice analysis. Even if someone else writes the address, the speaker must still engage in cognitive actions in speaking to their audience that may be evaluated based on thought. The critical question is, "Does he or she believe what she is saying?" Minute audiometric measurements can provide a clear picture of whether the speaker is deceptive. Someone who is disseminating information through speech but does not believe what they are saying is engaged in deception.

Let me provide you with a few examples, and I will describe the process in the following to evaluate the information provided.

Written Documents

Suppose someone provided me with a letter or written statement (exemplar) concerning an answer to something meaningful. My first step would be to read the document and understand its intent. It would be helpful if I could learn more about the individual who wrote the exemplar. Since I only speak and read English, the document in any other language would stop me cold. Looking for an interpreter probably would not help much because language structure, syntax, and other elements might not fall into the same investigative design I employ in SHIELD Analysis.

So, let us suppose that the document was in English. How would I proceed to evaluate its veracity?

Step 1:

As simple as it may seem, I thoroughly read the document. I look for apparent inconsistencies. Sometimes the deception is so evident that there is little need to go further. There may be evidence captured in a video, audio recordings or even print media that refutes points in the document. Politicians are famous for forgetting the power of video.

Step 2:

I determine whether the document is original or authored by another person used in a speech or public address. Evaluating speechwriters' works is a waste of energy because the language does not originate with the assessment subject. My efforts may diverge here. More about that later. If the document contains the original language, I move on to the next step.

Step 3:

I use a program called LIWC. That stands for Line Inquiry Word Count. It costs around $250 if you can still get it. It was created by James Pennebaker, Martha Francis, and Roger J. Booth back in 2001 at the University of Texas. The last I checked, there were requirements to gain access to the application based on research needs.

I developed a proprietary dictionary containing words and word phrases indicative of deception and installed it in LIWC. I take the text from the document I am examining and allow the software to conduct a word search from the dictionary terms. It highlights any dictionary terms within the paper a gives me a quick view of what may exist. Then I proceed to the next step.

Step 4:

Next, I employ a free program called TROPES. A trope is a figure of speech through which speakers or writers intend to express meanings of words differently than their literal meanings. It is a symbolic or figurative use of words in which writers shift from the literal meanings of words to their non-literal meanings. The trope could be a phrase, a comment, or an image used to create an artistic effect.

We may find its use almost anywhere, such as in literature, political rhetoric, and everyday speech. TROPES software is a perfect tool for Information Science, Market Research, Sociological Analysis, Scientific and Medical studies, and more.

Intended for use in content analysis, semantic classification, keyword extraction, linguistic and qualitative research, this application can parse words, phrases, or images. When combined with the LIWC product, a clearer picture emerges to find the presence of deception.

Step 5:

Now that a trail exists within the document of keywords and phrases, I begin the evaluation process. The process involves six modules of examination and an overall interpretation of the outcome using SHIELD Analysis. SHIELD is an acronym for Strategy, Heuristics, Identification, Emotional and Mental traits, Linguistics, and Demeanor.

I will not waste your time describing the process further because all the techniques are thoroughly explained in my book, Decoding Treachery, available on Amazon. When I complete the examination, I usually have an explainable and demonstrative product. I know whether the document represents the author in a deceptive role or a truthful one. In other words, is the author intentionally lying to me by providing falsehoods or concealing information to make me believe something that is not true.

Step 6:

When I find it relevant, as in President-elect Biden's case, there are occasions when I want to look at a person's language from a psychiatric point of view. In this instance, I employ software called PCAD3, developed by Dr. Louis A. Gottschalk, Ph.D., and Robert Bechtel, Ph.D. PCAD3 stands for "Psychiatric Content Analysis and Diagnosis, version 3." Based on a well-validated program called Content Analysis of Verbal Behavior, also by Dr. Gottschalk. This program looks at several exemplars of a person's speech. It provides a detailed view of different traits that may impact their behavior, including stress scales and psychiatric diagnosis.

Since I am not a licensed clinician, I refrain from diagnosing either a person's psychological or psychiatric traits. However, I find the tool extremely useful in explaining behavior that appears bazaar and subsequently only suggests possible reasons for the action based on the software's output.

Step 7:

This step is where my processes may diverge, as I explained in Step 2. More frequently than not, speeches and conversations are scripted, when the speaker is merely regurgitating an address on a teleprompter. When this occurs, any evaluation of the written format is a waste of time. However, it is essential to know what the speaker is thinking as he or she is speaking. There is a significant degree of scientific study that explains the relevance of speech to cognition.

Speech can be measured and correlated with specific thoughts. There are more than 129 parameters of measurement identified with the use of sophisticated software. In this instance, it is not the language that we evaluate but the human speech frequencies.

Without getting into the spaghetti of information on the subject, suffice to say that by using these software applications, I can determine emotional status, cognitive impairment, deception, malintent, stress, energy levels, concentration, compassion or empathy, and hundreds of other patterns.

Both software applications were developed by Nemesysco, Ltd, in Netanya, Israel. I was fortunate enough to have been thoroughly trained by their staff, and over the past several years, I developed expertise that classified me as an expert in their use.

The first application I utilize is called LVA7. LVA stands for "layered voice analysis." This program allows me to determine the emotional state of an individual during their speech. Using this application, I have been able to identify traits suggesting extreme pathologies of sociopathy, narcissism, and pathological liars. You might imagine the number of your political leaders who fall into these categories. With the advent of fake news, many so-called journalists fall equally into these categories as well.

The LVA7 also provides a statistical risk factor of potential deception and classifies a speaker into low, medium, and high-risk categories. What is important is that even when a speaker is determined to be low risk, there may still be deception. In one example, I evaluated House Speaker Nancy Pelosi's statement about her piousness and devout religious principles. Her risk factor was about 31%. Although she presented an exceptionally low-risk factor, the specifics within that 31% provided more facts. That is what Paul Harvey referred to as "the rest of the story."

The second application I use is called LVA 6.50. LVA 6.50 is an investigative software tightly controlled by the Israeli Ministry of Defense. They license it to individuals of friendly governments and countries determined to be non-detrimental to Israel's security interests. The software is neither secret nor classified, but they control it aggressively.

This software allows me to drill down into specific deception patterns and look for statistical anomalies within a subject's speech. Because the application is so thorough and intense, I use it as the last process in my evaluations. This application is so explicit that I could identify when the fly landed on Vice President Pence's hair during his debate with Kamala Harris and how long it impacted his responses. I am almost sure the fly was Russian, helping Vice President Pence win the discussion. Lookout, another left-wing Russian conspiracy on the make. It was detected by the LVA 6.50! Not really…just kidding.

Summary

Now you have a clearer understanding of the processes I use to evaluate the integrity of communications from speakers in the public domain. For as long as I have been doing this, I am continually amazed at what I find. In nearly all political discourse cases, public benefit is secondary, and the primary interest is self-promotion and enhancement.

Even when the deception is blatant, many general populations are too susceptible to falsehoods. They become angry when they learn of the scam (some hold on to their beliefs regardless), and they are dismayed. Questioning authority was once the American way, but since the advent of social platforms and fake news, it is easier to remain a child and follow the pied piper's music.

In the preceding chapters, I provided you with examples of evaluations I performed past and present. I did not intend for them to be derogatory, but occasionally I cannot refrain from the obvious irony that results, and I must insert my dry sense of humor. Please forgive my foibles and evaluate for yourself. I do not play favorites. Political affiliation is secondary to me. I simply hate liars and do not hesitate to call out what I find.

Conclusion

Throughout this book, my goal was to expose you to what lying is all about and provide you with tools to defend against it. The average person does not usually have the skill or the high-powered software to screen information to determine its legitimacy.

While intuition is a great equalizer, it does not always provide enough insight to articulate a person's uneasiness when confronted with a potential lie. Consequently, one needs a tool he or she can rely on to provide some consistency and validation to internal feelings about potentially false information.

SHIELD Analysis is a language assessment technique designed to help you sort fact from fiction. Is it 100% accurate? It is not! But neither is your intuition or any other form of lie detection.

We try for the 100% level, and the more you practice with SHIELD Analysis, the better you will get. You will develop your assessment style, and I believe you will be amazed at the deception you recognize even in the very beginning.

Once you familiarize yourself with the processes, you may wish to advance to higher levels of training. Eventually, you may aspire to provide services to others, as I have done for nearly fifteen years. As you require more training, Western Legends Research is ready to offer intermediate and advanced levels. You can find the Basic SHIELD Analysis Course in my book, <u>DECODING TREACHERY</u>, along with a few extra tips.

The Intermediate SHIELD Analysis Course provides the next step in advancement. This intermediate course provides you with an opportunity to work through examples and exercises to practice what you learned. If you miss something, you can go back to this book or review the video to reinforce the concepts.

Eventually, maybe not so far in the future, you may wish to pursue the Advanced SHIELD Analysis Course. The advanced course is the full CMD diagnostics program where you will learn the uses and gain access to sophisticated software platforms like PCAD3, LIWC, TROPES, LVA-4i, and LVA 6.50.

These applications can provide you with answers in minutes rather than hours of in-depth analysis. At this point, you will have the skill to become a full-time practitioner or even an instructor. Can you envision yourself as the expert people come to with concerns before and after being the victims of deception?

Your new knowledge allows you to not only discover the deception, but you now have the skill to teach the techniques to others. These are the same techniques and software applications I use to reduce my time in language and speech assessment.

Take your time with the learning process, and whether you learn by reading or by viewing videos, I will teach you what you need to know. Honestly, this does not take a Ph.D. in some social science. I have introduced these techniques to young people right out of high school and older adults well into their seventies.

The concepts are not complicated, and once you learn them, they will stay with you forever.

You will become more attentive to conversations and written communications. You will learn to listen to people instead of just hearing them.

I hope you have enjoyed the book and found some benefit. I look forward to working with you to pursue higher levels of training. And as you advance, feel free to contact me anytime at:

WESTERN LEGENDS RESEARCH, LLC

P.O. Box 5343

Codename: Language Inquisitor

Sun City West, AZ 85376

Email: dtunnell@westernlegendsresearch.com

Web: https://www.westernlegendsresearch.com

Thank you for reading.

Dale Tunnell

Dale Tunnell, Ph.D.

About the Author

DALE TUNNELL was born in Powell, Wyoming, in 1951. He is a decorated Vietnam veteran, married and now living a retired lifestyle in Phoenix, Arizona.

Trained in psycholinguistics and psychological content analysis, Dale is a retired law enforcement officer with over forty years of service with federal, state, and local agencies.

He earned his Master of Arts Degree in Management from Webster University and his Doctor of Philosophy Degree in Psychology from Capella University. Dale received Beginning, Advanced, and Stage II training In Scientific Content Analysis at the Laboratory for Scientific Interrogation, in Phoenix, Arizona. He also mentored under Louis Gottschalk, MD, Ph.D., at the University of California at Irvine, where he acquired his psychiatric content analysis and diagnosis expertise.

Dale served as a Senior Researcher for Nemesysco, Ltd, Netanya, Israel, and is recognized internationally as an expert in Layered Voice Analysis. He was also the Director of Forensic Intelligence and Research with Halcyon Group International.

He is an author of a previous book about the secrets William H. Bonney, alias Billy the Kid, nearly took with him to his grave. In his book, **RESURRECTING THE DEAD,** he used many of the methods described in this current book. Find all my books under my name on Amazon.com and numerous other booksellers worldwide.

In another book, **DECODING TREACHERY,** he teaches the skills he uses to identify deception and unmask a liar's language.

RESURRECTING THE DEAD and ***DECODING TREACHERY*** are available on Amazon.

Be sure and keep an out for his future books:

CODENAME: The CHINA DOLL

CODENAME: The FLYSWATTER

CODENAME: The GIGGLER

CODENAME: The GRIFTER

CODENAME: The WANNABE

CODENAME: The KENYAN

Dale is an active member of the American Psychological Association and the Linguistic Society of America.

Made in the USA
Columbia, SC
16 January 2021